EXPLORING

HANAUMA

BAY

Text by
Susan Scott

Photographs by
David R. Schrichte

A KOLOWALU BOOK

UNIVERSITY OF HAWAII PRESS

HONOLULU

© 1993 University of Hawaii Press
All rights reserved
Printed in Singapore

98 97 96 95 94 93 5 4 3 2 1

Designed by Paula Newcomb

Library of Congress Cataloging-in-Publication Data

Scott, Susan, 1948–
    Exploring Hanauma Bay / text by Susan Scott ; photos by David Schrichte.
       p. cm.
    Includes index.
    ISBN 0-8248-1494-0
    1. Skin diving—Hawaii—Oahu—Guidebooks. 2. Scuba diving—Hawaii—
Oahu—Guidebooks. 3. Hiking—Hawaii—Oahu—Guidebooks.
4. Oahu (Hawaii)—Guidebooks. I. Title.
GV840.S78S35   1993
797'.09969'3—dc20
93-26657
CIP

CONTENTS

ACKNOWLEDGMENTS

This book is the result of many people's efforts. We thank George Balazs, Robin Bond, Larry Braver, Bruce Carlson, Ann Fielding, Friends of Hanauma Bay, Michael Hadfield, Robert Harman, Terry Heckman, Alan Hong, Cynthia Hunter, Lisa King, Craig MacDonald, Kathleen Mahannah, John Randall, Ernest Reese, Sea Life Park's Education Department, Guy Tamashiro, Leighton Taylor, Craig Thomas, Robert Titgen, Underwater Photographic Society, Don Vehon, Waikiki Aquarium Staff, George Walker, and Thomas Zaleski.

INTRODUCTION

We wrote *Exploring Hanauma Bay* for walkers, waders, snorkelers, and divers of all ages. To cover such wide ranges of abilities and interests, we divided our book into six tours, from walking to scuba diving. Each tour begins with a map of the suggested route, then photographs show what you might expect to see there. Since most of these creatures are constantly on the move, it's possible to see the same kind of fish or invertebrate on several tours.

No one knows how many different kinds of plants and animals now live in Hanauma Bay. We have featured most of the common animals and several rare ones, but it's possible you'll spot more. This chance of discovery is just one of the many charms this nature park offers.

Every photograph in this book was taken in Hanauma Bay. Since the area was nearly barren in 1967 from overfishing, these photographs are dramatic evidence of the effectiveness of creating marine life conservation districts. We hope that this guide helps people enjoy and take care of Hanauma Bay and that it will inspire the creation of more underwater parks in Hawai'i.

DO'S AND DON'TS OF HANAUMA BAY

DO:

- Check with lifeguards about present conditions.
- Avoid swimming in strong currents. If caught in one, stay calm. Wave and shout for help.
- Watch the waves while walking on ledges and move back if one comes toward you.
- Use waterproof sunscreen. Other types wash off, clouding the water.
- Use restrooms. Human urine pollutes bay waters.
- Pick up trash.
- Respect the animals that live here by not touching or chasing them.

DON'T:

- Stand on or grab hold of any coral. This injures the coral animals.
- Rely on flotation devices if you can't swim.
- Walk on wet rocks on ledges. These places are slippery and show you that waves are breaking there.
- Feed fish peas, bread, or other human food.
- Take anything natural from the bay, including sand, plants, or rocks.
- Approach, touch, or try to ride sea turtles, dolphins, or manta rays. These activities are illegal and can hurt the animals. If they come to you, hold still and enjoy the experience.

WEIGHTS AND MEASURES

U.S. Unit	Metric Equivalent
1 pound	0.373 kilograms
1 inch	2.54 centimeters
1 foot (12 inches)	30.48 centimeters

TOUR 1: WALKING

Toilet Bowl

Witches' Brew

Telephone Cable Channel

Back Door

FOR

Everyone who enjoys a shoreline walk.

Because waves that break onto ledges can sometimes cause injury and even death, walk these areas with extreme caution. If in doubt about surf conditions, check with lifeguards. (See Safety Tips below.)

EQUIPMENT

Wear shoes. The sand gets hot, and the lava rock is sharp in places. Tough feet, however, can make this tour barefoot.

Polarized sunglasses aren't essential, but they do reduce the glare on the water's surface, making it easier to see and identify fish in the water.

DIRECTIONS

To explore the Toilet Bowl side of the bay, turn left as you face the bay from the sand beach. The area here is wide and full of pools created by waves splashing onto the ledges during high surf and high tides. As you continue out, a rock slide crosses the path, involving some easy rock scrambling. Beyond that patch, the walking is smooth and flat all the way to the Toilet Bowl.

Witches' Brew, named for the turbulent water in this area, is located at the first point on the opposite side of the bay. Facing the bay, turn right, walk to the end of the beach, then head out. The ledge walk on this side is easy until you come to a 1991 rock slide that blocks the path. Be careful here. The scramble around these fallen rocks takes you dangerously close to the water's edge and breaking waves.

SAFETY TIPS

When walking near the water's edge, you can tell where the waves are breaking by noticing wet places on the rocks. But because wave heights change with the winds and tides, these wet places are only rough guidelines for walking. Occasionally, tragedy strikes Hanauma Bay when a surprisingly big wave

Parents should accompany children on all ledge walks.

Enjoy the flushing water of the Toilet Bowl from the safety of the dry ground around it. Watch your step near the edge. Lifeguards are routinely called to help people who have slipped on the algae-covered rocks around the bowl.

Witches' Brew during high surf. To stay safe, keep well away from all wet spots on ledges.

breaks over the path and takes someone by surprise.

Waves arrive in groups called *sets* followed by quiet periods called *lulls*. When you're near wet places on ledges, watch the waves for a few minutes (not seconds) to feel this rhythm. Cross between sets, during lulls.

Because most newcomers can't spot a big wave in time to get out of the way, it's safest to simply stay away from all wet places on the ledges. But you can still get caught. If you see a big wave coming your way, run toward the cliff wall immediately.

The currents that rush in and out of this geological oddity called the Toilet Bowl sometimes injure and even kill people who jump into the bowl for rides.

3

The Making of Hanauma Bay. One of several volcanic vents called the Honolulu Volcanics created what is now Hanauma Bay. During the eruptions, seawater mixed with lava, building a glassy ash ring around the vent. Eventually, chemical changes turned the ash to rock. Over the last 30,000 years, waves, wind, and rain have sculpted Hanauma's crater walls into the sea cliffs that line the bay today.

Ancient Reef Limestone. During eruptions at Hanauma, steam and lava explosions sometimes flung out pieces of ancient reef limestone that broke off vent walls. The white rock in this photo is likely one of those pieces.

Sea Lettuce *(limu pālaha-laha), Ulva fasciata.* This bright green seaweed is one of the most common types in Hawai'i. Look for it in the bay's tide pools and on the rocks between areas of high and low tides. In deeper water, this seaweed is food for turtles and fish. Hula dancers in ancient Hawai'i used sea lettuce as adornment. The Hawaiian name for all sea-weeds is *limu,* a word commonly used in Hawai'i today.

Turtle or Holly Limu *(limu kala), Sargassum echinocarpum.* This is another common seaweed in Hawai'i, growing in both rocky and sandy areas. The brownish leaves are often jagged, like holly, and bear conspicuous dots. Chubs *(nenue),* bluespine unicorn-fish *(kala),* and green sea turtles *(honu)* are among the marine animals that eat this type of seaweed.

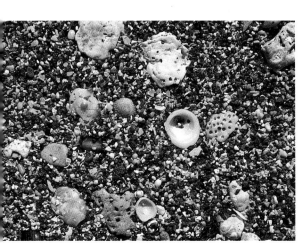

Sand. Hawai'i's white sand is made of marine animal skeletons and pieces of calcium-containing seaweeds. Look closely at a handful of sand for crab legs, snail shells, urchin spines, coral, or other parts of marine animal skeletons. The bay's darker sand, near the Witches' Brew side of the bay and at the Toilet Bowl area, contains bits of black lava and a sparkling, greenish mineral called olivine.

Ghost Crab *('ōhiki), Ocypode ceratophthalmus.* Female. As you walk along the beach toward Witches' Brew, look for holes in the sand. These are ghost crab burrows. The shy, sand-colored crabs that live there are most active at night, scavenging the beach for bits of animal material that has washed in. Sometimes, however, ghost crabs like this one will peek out of their holes during the day. Ghost crab shells reach about 3 inches across.

Ghost Crab Holes. You can learn a lot about a ghost crab by looking at its home. The size of the hole is the size of the crab that lives there. Males build well-formed sand mounds a few inches from their holes, but females and juveniles pile their discarded sand in loose heaps or throw it in wide fans.

Rock Crab *('a'ama), Grapsus tenuicrustatus.* The low bodies and long, spiny legs of rock crabs enable them to hang onto wave-washed rocks. Look for these dark crabs at the water's edge, where they graze on seaweed. People at traditional Hawaiian lū'aus often eat rock crabs salted and un-cooked, and anglers sometimes use these crabs for bait. Rock crab shells reach about 3 inches across.

Periwinkle *(pūpū kōlea), Littorina* sp. These spiral-shaped, 1/2-inch-long snails live above the high-tide line, where they graze on seaweed and tissue from decomposing plants and animals. During dry periods, periwinkles keep from dehydrating by closing their patterned gray-and-tan shells tight, then sticking to the rocks with mucus. It's best to look but not touch these or any other intertidal animals because they spend significant energy getting to and staying in certain feeding zones.

Rock Crab Skeleton. This is not a dead rock crab but the molted shell of a crab that outgrew its external skeleton. The shell is thin and fragile because the animal absorbed much of the calcium into its new shell before it backed out of the old.

Nerite *(pipipi), Nerita picea.* About the size of a fingernail, these blackish blue snails live in crevices near the water's edge, where they graze on seaweed. This kind of snail is abundant in Hawaiʻi and Johnston Island but is rare everywhere else in the Pacific.

Shingle Urchin *(hā'uke'uke),*
Colobocentrotus atratus. Shingle
urchins are well adapted for survival
in their wave-washed environment.
Hundreds of suctioning tube feet hold
these black seaweed grazers firmly
to rocks while the animals' smooth,
low profiles resist the force of the
surf. The spines of shingle urchins
fit together like tiles, protecting the
animals from dehydration during low
tides. Shingle urchins grow to about
3 inches across.

Hermit Crab *(unauna), Calcinus*
seurati. Hermit crabs never kill snails
to get their shells but move into already
empty ones. The hard snail shells
protect these crabs' soft, vulnerable
abdomens. Look in tide pools and
along shallow, rocky shores for these
1/2-inch-long, blue-eyed hermits that
eat decomposing plant and animal tissue. Because they will die out of water,
it's important to return hermit crabs to
their pools when you've finished looking at them.

Zebra Blenny *(pāo'o),*
Istiblennius zebra. If you see
dark, narrow fish darting around
tide pools, they're probably
zebra blennies. This blenny is
found only in Hawai'i, but close
relatives occur throughout the
tropical Pacific Ocean. These
hardy fish can leap from pool to
pool over dry ground, surviving
great changes in temperature,
salinity, and oxygen concentra-
tion. Zebra blennies grow to
about 7 inches long and come in
several colors, from nearly black
to light tan with dark bars on
the body. They eat bottom-growing seaweed and decomposing
plant and animal tissue.

Achilles Tang *(pāku'iku'i), Acanthurus achilles.* As you walk toward the points on both sides of the bay, look for bright flashes of orange in the surging water near the ledges. These are achilles tangs, a type of surgeonfish. Several kinds of surgeonfish, also called tangs, are common throughout the bay. All have sharp spines ("scalpels") on each side at the rear of the body. With a swish of the tail, these sharp spines can injure intruding fish. Achilles tangs advertise their scalpels with bright orange spots against black bodies. These aggressive seaweed grazers chase away other fish that wander into their grazing space. Achilles tangs grow up to 10 inches long.

Pigeon, *Columba livia.* The tame birds so common on the beach and in the park's cliffs are common pigeons, also called rock doves, introduced to Hawai'i years ago and now thriving. These birds have many color variations, including white, gray, brown, and mixtures of these. Because of over-population problems, it is now illegal to feed these charming beggars, which often carry diseases transmittable to humans. Hanauma Bay's pigeons are smart; hide all food and anything that resembles food before leaving for a swim or walk. Pigeons grow to about 12 inches long.

Pacific Golden Plover *(kōlea), Pluvialis fulva*. Plovers are 11-inch-long migratory shorebirds that winter in tropical islands. In spring, most golden plovers fly to the Arctic, where they mate and raise young. When these chores are finished, the birds return to the tropics, making the long, nonstop flight in just a few days. Look for Hawai'i's golden plovers from August through April. As spring approaches, the drab brownish feathers of these birds gradually change to bold breeding colors of black, white, and golden brown. These migrants are common in the grassy areas of the park, where they run in short bursts and then stab the ground to catch insects and other invertebrates.

Mongoose *('iole manakuke), Herpestes auropunctatus*. In 1883, sugar planters imported the first mongooses to Hawai'i from India via Jamaica to control cane-field rats. However, mongooses sleep at night, a habit that limits their success in catching nocturnal rodents. Mongooses eat just about anything, including lizards, crabs, birds, eggs, tide-pool fish, and fruit. You'll most likely see these brown pests around trash containers in the park, where they scavenge for human food. Adult mongooses weigh from 1 to 3 pounds and measure from about 1-1/2 to 2 feet long (including the tails).

Common Mynah, *Acridotheres tristis.*
Dr. William Hillebrand introduced these
9-inch-long, black-and-yellow birds to
Hawai'i from India in 1865 to eat crop-
destroying armyworms. The birds are
now common throughout the Islands,
including the grassy areas and trees of
Hanauma Bay. Mynah birds continually
entertain people with their strutting
walk and noisy calls.

English Sparrow, *Passer domesticus.*
These 6-inch-long grayish birds are common throughout the park as well as on
all the main Islands. English sparrows,
which eat practically anything, were
introduced to Hawai'i from New
Zealand in 1871, and then flourished.
It's easy to spot sparrows and their
nests throughout the grassy areas of the
park and in the rock cliffs.

Red-crested Cardinal,
Paroaria coronata. Also
called the Brazilian cardinal,
this 7-inch-long redhead with
the gray body was introduced
to O'ahu from South America
in the 1930s. It is now common in parks on O'ahu as
well as on Kaua'i, Moloka'i,
and Lāna'i. Both sexes of
these cardinals look alike.
Immature birds have brown
heads, which later turn red.
Red-crested cardinals eat
seeds, plants, insects, and
fruit.

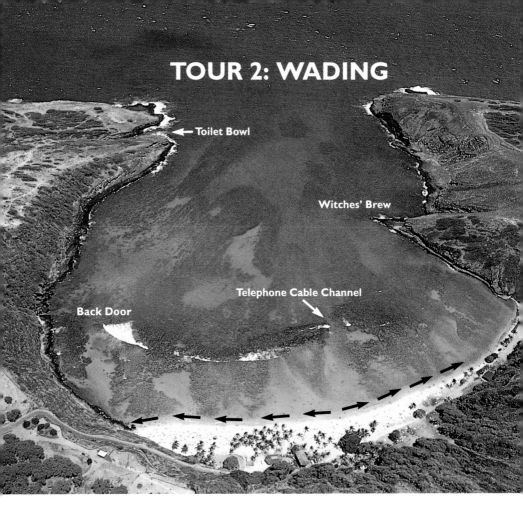

TOUR 2: WADING

Toilet Bowl

Witches' Brew

Telephone Cable Channel

Back Door

FOR

Everyone who enjoys wading with fish.

EQUIPMENT

Polarized sunglasses aren't essential but will help you see the fish better by reducing glare on the water's surface.

DIRECTIONS

Wade in the sandy areas near the center of the beach. Low-tide mornings offer the best views because the water is clear and calm then. Walk into the water only as far as you feel comfort-able, then look down. Fish are usually nearby, often in water just a few inches deep.

ABOUT FEEDING FISH

It was once common practice to bring frozen peas or bread to feed Hanauma Bay's fish, but no more. Biologists believe that these and other human foods hurt the animals that you've come to admire. The concession stand now sells fish food designed to keep the fish healthy.

SAFETY TIPS

Nonswimmers and parents with young children should wade out only as far as they feel comfortable. Usually you don't have to go beyond ankle-deep water to see schools of fish.

During feeding, some fish can get confused and accidentally nip a leg or toe. To avoid this, toss fish food away from yourself and others.

At Hanauma Bay, you don't have to feed the fish to get a good look at them. If you do offer food, however, keep the fish healthy by giving them only approved fish food.

Both children and adults enjoy the charm of this underwater park where marine animals and people meet. The bluish gray fish here are chubs, also called rudderfish.

Hawaiian Flagtail *(āholehole), Kuhlia sandvicensis.* These 6- to 12-inch-long fish are among the most common you'll see in the bay's shallow water. Flagtails are silver, with big eyes and forked tails. Their backs are slightly darker than their sides. Hawaiian flagtails, native only to Hawai'i, are usually nocturnal but here they come out in large numbers during the day to eat offered fish food. Normally, these fish eat crabs and shrimps adrift in ocean currents, worms, and seaweed.

Mullet *('ama'ama), Chaenomugil leuciscus.* Most mature mullet are about 20 inches long, but because of the bay's abundance of food, the mullet here sometimes grow even larger. Mullet are extremely common in the shallows. Look for bulky, silvery fish with rounded snouts and downturned lips. These fish normally cruise over the bottom eating decomposing plant and animal material.

Chub *(nenue), Kyphosus* sp. Chubs, also known as rudderfish, usually live on coral reefs or in places with rocky bottoms. Although they normally graze on bottom-growing seaweed, at Hanauma Bay these fish are among the first to show up for handouts in both shallow and deep water. Mature chubs are larger than mullet, growing to about 24 inches long. You can pick the silvery chub out of a crowd by looking for fish with a bluish tinge.

Spotted Trunkfish
(moa), Ostracion melea-gris. Female. Female and juvenile spotted trunkfish have white spots over dark brown bodies. Both sexes of spotted trunkfish in Hawai'i are a slightly different color than those in other parts of the Pacific. Spotted trunkfish normally eat sponges, seaweed, and sea squirts, but those in Hanauma Bay frequently show up for handouts.

Spotted Trunkfish *(moa), Ostracion meleagris.* Male. These well-named fish look like 6-inch-long trunks with fins. Males have brown backs with white spots, and blue sides with black spots. Their boxy skeletons act like suits of armor, making them hard for predators to swallow. Another defense that all trunkfish possess is the ability to secrete a toxin, which also discourages predators.

Yellowfin Surgeonfish *(pualu), Acanthurus xanthopterus.* At 22 inches long, these are the largest of Hawai'i's surgeonfish. Yellowfin surgeonfish are brownish gray with pale yellow markings. These surgeonfish range throughout the Indian and Pacific oceans and can be found in harbors, bays, near coral reefs, and sometimes in deep water far from the shelter of the reefs. Yellowfin surgeonfish eat most anything they find in the sand and also swallow some of the sand itself, which probably helps the fish digest seaweed.

Milletseed Butterflyfish *(lauwiliwili), Chaetodon miliaris.* These 6- to 7-inch-long fish, found only in Hawai'i, are Hawai'i's most common butterflyfish. Milletseed butterflyfish range from shorelines to beyond scuba limits (around 100 feet), eating tiny, drifting animals. Their yellow-and-black-spotted, vertical pancake shapes are unmistakable in the bay's shallow waters.

Convict Tang *(manini), Acanthurus triostegus.* Convict tangs often graze in large groups that can push away lone, seaweed-eating fish. In the bay, these pale greenish gray fish with black bars come in from the reef in small groups to eat fish food. Look in pools along the sides of the bay for inch-long juveniles. Adults grow to about 10 inches long.

Christmas Wrasse *('āwela), Thalassoma trilobatum.* Male. It's easy to spot these blue, red, and green fish through several feet of shallow water. Normally, Christmas wrasses, named for their red and green colors, are found in rocky-bottomed surge zones, but in the bay, they often show up at feeding sessions. These 11-inch-long fish normally eat crabs, snails, and small sea star (starfish) relatives called brittle stars.

Yellowstripe Goatfish *(weke)*, *Mulloides flavolineatus.* All goatfish have two chin whiskers, called barbels, covered with taste buds. As the fish cruise over sandy bottoms, they wiggle their barbels in the sand to find crabs, shrimps, other invertebrates, and fish. You can easily see goatfish stirring up sand in the shallow water of the bay. Yellowstripe goatfish, which grow to about 16 inches long, blend well against sandy backgrounds. Look for a black spot and a pale yellow stripe on each cream-colored side for identification.

Manybar Goatfish *(moano)*, *Parupeneus multifasciatus.* Manybar goatfish are purplish red and, at 7 to 11 inches long, are smaller than yellowstripe goatfish. Look for dark bars across the backs and down the sides of these fish. Manybar goatfish are found from the shallow waters near the shore to beyond scuba limits, digging with their whiskers for crabs, shrimps, snails, octopuses, fish, and fish eggs. In the photo, a Hawaiian cleaner wrasse picks parasites and scales from the skin of a manybar goatfish.

Hawaiian Sergeant *(mamo), Abudefduf abdominalis.* These light blue-green fish with black bars belong to the damselfish family, whose members are known for aggressively defending their territory even against human intruders. These 6- to 9-inch-long fish don't bite. However, males defending eggs, the purple masses below the fish in the picture, will zoom close to a swimmer or diver. Sergeants mostly eat tiny drifting marine animals, but they also graze on seaweed. A few of these common fish usually show up in shallow water during feeding sessions.

Blackspot Sergeant *(kū-pīpī), Abudefduf sordidus.* These yellowish gray relatives of Hawaiian sergeants are easily identified by the black spot ahead of the tail fin. Blackspot sergeants are found along shallow rocky bottoms in surge areas where the water moves back and forth with the waves. These fish eat just about anything, including seaweed, crabs, barnacles, sponges, and worms. Blackspot sergeants grow to just over 9 inches long.

Bullethead Parrotfish *(uhu), Scarus sordidus.* Male. Parrotfish use their beaklike teeth to graze on seaweed growing in and on dead coral. Often, the fish scrape off some coral rock along with the plants, grinding the two in their throats. The fish absorb the nutrients but pass the bits of coral skeleton through their digestive tract. Through this process, parrotfish are major producers of sand on tropical reefs. Bulletheads grow to just over 15 inches long.

Bullethead Parrotfish *(uhu), Scarus sordidus.* Female. Parrotfish are sometimes hard to identify because juveniles, males, and females are often different colors. To make identification even more difficult, females frequently turn into males, gradually changing colors as they do so. In general, female parrotfish, like this bullethead, are drab gray to reddish brown, and males are shades of blue and green.

Female Parrotfish in Cocoon. All parrotfish sleep at night among the reef's rocks and coral. Some, like this one, secrete a mucous cocoon around themselves, perhaps to mask their smell from predators.

Stareye Parrotfish *(pānuhunuhu),* *Calotomus carolinus.* Female. Stareye parrotfish get their common name from the males because only males have pink bands radiating from their eyes. Both males and females of this type grow to about 20 inches long.

TOUR 3: BEGINNING SNORKELING

Toilet Bowl

Witches' Brew

Telephone Cable Channel

Back Door

FOR

Swimmers of all ages.

EQUIPMENT

You'll need a mask and snorkel. Fins help you swim with less effort but are optional. If you don't have your own snorkeling gear, you can rent it at the park. It's important to get the right sizes. If your mask leaks or fins don't fit, try adjusting the straps or ask for replacements. Water-filled masks or aching feet can cloud this fine snorkeling experience.

It's common to see experienced snorkelers and divers spit into their masks. This looks silly but the coating prevents masks from fogging, another annoyance that can put a damper on the day.

DIRECTIONS

Enter the water anywhere from the sand beach and head toward one of the open spaces inside the reef. If the tide is high, you can swim over the top of the reef flats to adjacent open spaces. During low tide, however, you'll have to come back to the beach, walk to a new area, and then head out again.

SAFETY TIPS

Always swim with a buddy and tell him or her if you're having trouble with your equipment or are feeling uneasy. If you're more comfortable in shallow water, snorkel there.

Beginners should stay away from the two channels leading outside the reef. These are labeled Telephone Cable Channel and Back Door on the map. Both channels can have outgoing currents that can funnel you into deep water. If you do get caught in a current and can't get back, stay calm, then call or wave for help. Lifeguards watch these areas.

Be careful where you put your hands and feet. Grabbing coral not only damages it but can cut you. Also, coral reef holes are the homes of several types of marine animals that might bite if you suddenly stick a hand or foot in their faces. Rest in the sandy areas.

You don't have to swim far in this teeming sanctuary to see plenty of fish. The milletseed butterflyfish shown here are the most common butterflyfish in Hawai'i.

Coral. Most of the coral on the inside of the reef is dead, killed by people walking on it and grabbing hold of it. Dead coral is white or brownish; living coral contains shades of green, blue, pink, or yellow. These colors come from tiny plants that live within the tissues of all reef-building corals. The plants also provide nourishment. Never walk, stand on, or hold onto any coral. This only hurts it and prevents recovery of the reef.

Rock-boring Urchin *('ina uli), Echinometra mathaei.* Notice the common, round holes about 2 inches in diameter in the coral rock here. Rock-boring urchins make these holes, probably by a combination of chemical action and scraping at the coral with their teeth and spines. Most of these urchins spend their entire lives in their burrows, living on bits of seaweed that get caught on their spines. These spines are only mildly sharp and contain no toxin. Some rock-boring urchins are black, but this particular kind is usually pale pink or green tinged.

24

Thornback Cowfish *(makukana), Lactoria fornasini.* These unusual, 5- to 6-inch-long fish are members of the trunkfish family, which look like little trunks with eyes, mouth, and fins. Cowfish paddle slowly around rocky and sandy bottoms, giving snorkelers good viewing opportunities. The common name of these tan fish with blue spots probably came from the "horns" on their heads.

Hawaiian Whitespotted Toby, *Canthigaster jactator.* Tobies belong to the pufferfish (or blowfish) group of fish that, when threatened, can inflate themselves by sucking air or water into their bodies. Another feature common to most pufferfish is the presence of a powerful toxin in their skin, flesh, and organs. The brown, 3-inch-long whitespotted toby, found only in Hawai'i, is both the smallest and the most common of Hawai'i's pufferfish. These fish eat a wide range of plants and animals.

Hawaiian Cleaner Wrasse, *Labroides phthirophagus. (Top)* This 4-inch-long purple-and-yellow wrasse, which lives only in Hawai'i, picks parasites, scales, and mucus off other fish. The picture shows a cleaner wrasse with its saddle wrasse relative.

Saddle Wrasse *(hīnālea lauwili), Thalassoma duperrey. (Bottom)* This 10-inch-long wrasse, another fish found only in Hawai'i, is the most abundant reef fish here. Saddle wrasses eat small invertebrates such as crabs, worms, snails, and urchins, as well as decomposing animal tissue. You can see these common saddle wrasses from several feet of water to about 70 feet deep. Look for a blue-green fish with an orange "saddle" just behind the head.

Hawaiian Cleaner Wrasse and Surgeonfish. Cleaner wrasses swim in specific areas called cleaning stations that larger fish visit, holding still while the cleaner wrasse does its work. Some researchers believe that fish line up for this service not to get rid of parasites but to get a rub from the fins of the cleaner wrasse. This little cleaner gets a meal from an orangeband surgeonfish in exchange for a back scratch.

Hawaiian Cleaner Wrasse and Whitemouth Moray Eel. An eel opens wide to get its teeth cleaned by this cleaner wrasse. Look for wrasse cleaning stations in small open spaces close to the reef.

Bird Wrasse *('akilolo), Gomphosus varius.* Female. These unmistakable wrasses with the long "beaks" grow to about 12 inches long. Like most other wrasses, male, female, and juvenile bird wrasses have different coloration: males are dark blue-green, females (pictured) are black with white heads, juveniles have dark green backs with white bellies. All bird wrasses, however, have unusually long snouts that make these fish easy to identify. Bird wrasses eat mostly bottom-dwelling crabs and shrimps.

Blacktail Wrasse *(hīnālea luahine), Thalassoma ballieui.* The blacktail wrasse, native only to Hawai'i, is one of the largest in the wrasse family, growing up to 2 feet long. The Hawaiian name means old lady wrasse, so-named because of the grayish color of the body. Older blacktail wrasses have checkerboard patterns on their sides. These fish eat mostly sea urchins, crabs, and fish.

Eightline Wrasse, *Pseudocheilinus octotaenia.* This 5-inch-long wrasse is a common shallow-reef inhabitant, but its shy nature often makes it hard to spot. Look for a splash of orange darting into the rocks. Eightline wrasses, which have eight dark stripes on each side, mostly eat crabs and shrimps from the ocean floor.

Ornate Wrasse *('ōhua), Halichoeres ornatissimus.* These fish in the brilliant red, blue, and green outfits are common in the shallow waters of the bay. Ornate wrasses, which grow to about 8 inches long, eat small crabs and snails. To distinguish ornate wrasses from their brightly colored wrasse relatives, look for a black spot behind each eye.

Scarface Blenny, *Cirripectes vanderbilti.* The little tentacles above the eyes of these 4-inch-long blennies look like eyelashes, giving these fish endearing expressions. Orange markings around the eyes and face of this blenny help with identification. Look for these common brown blennies perched on coral rocks from 2 to 30 feet deep. Scarface blennies eat seaweed and decomposing plant and animal tissue.

Moorish Idol *(kihikihi), Zanclus cornutus.* Because of the long, streaming fin on its back, people often mistake the black, white, and yellow moorish idol for an angelfish. The two are not related. Moorish idols have the distinction of being the only member of their family. These graceful-looking fish, which grow to about 8 inches long, are found in both shallow and deep water. They eat several kinds of plants and animals, but their most common food is sponges.

Hawaiian Hogfish *('a'awa), Bodianus bilunulatus.* Juvenile. Adult hogfish are usually found in water over 25 feet deep, but you can see juveniles inside the reef at Hanauma Bay. These members of the wrasse family can grow to 20 inches long. Young fish, like the one pictured, are bright white and yellow with dark brown lines. Colors of adult female hogfish are similar but faded; males are deep purple. Hogfish eat mostly snails, sea urchins, and crabs.

Hawaiian Dascyllus *('ālo'ilo'i), Dascyllus albisella.* These 5-inch-long beauties are members of the damselfish family. A white spot on each side of the black fish expands to cover the whole fish when it ventures from shelter. It's common to find groups of *Dascyllus* hovering in and around antler and cauliflower coral heads, where the fish quickly dive for cover when startled. Hawaiian *Dascyllus* eat tiny drifting animals.

29

Devil Scorpionfish *(nohu 'omakaha), Scorpaenopsis diabolus.* Because these masters of disguise match their surroundings so perfectly, it's possible to be just inches from this fish and still not see it. Devil scorpionfish, which grow to about 12 inches long, flash the bright orange and yellow undersides of their front fins when startled. These are warning flags. Devil scorpionfish are not aggressive, but if you accidentally grab or step on one, their venomous spines can cause a painful wound. These and all scorpionfish eat crabs, shrimps, and fish that pass within striking distance.

Pacific Gregory, *Stegastes fasciolatus.* Gregories are damselfish that are usually found in quiet waters inside surge zones. These loners are among the more common shallow-water reef fish in Hawai'i. Look in reef cracks for these 5-inch-long, brownish gray fish with bright yellow eyes. Pacific gregories graze on seaweed and decomposing plant and animal material.

Orangespine Unicornfish *(umaumalei), Naso lituratus.* These 18-inch-long, brown unicornfish have no horn, but their brilliant orange-and-yellow markings make them easy to identify. Orangespine unicornfish graze on leafy, brown seaweed. Their common name refers to the color surrounding their scalpel-like tail spines. These large surgeonfish sometimes search for food in small groups in the shallow waters of the bay.

Manyray Flatfish *(pāki'i), Bothus mancus.* Because these members of the left-eyed flounder family change color to match their background, it's hard to see these fish unless they move. When they do, you won't forget the sight. These fish wiggle along the bottom like 19-inch-long flying carpets, changing color as they move. Manyray flatfish commonly rest on or partially under the sand inside the reef.

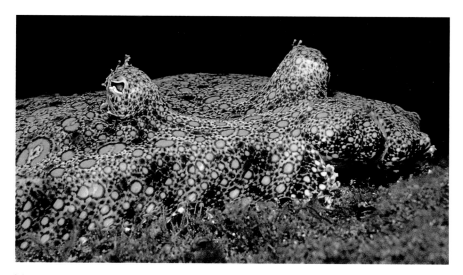

Manyray Flatfish Eyes. Immature flatfish look like normal fish, but as they mature, their bodies become asymmetrical. Because one side of the head grows faster than the other, the eyes appear to shift position—one moves across the top of the head to join the other eye, where both sit like gleaming marbles. The blind side of the fish becomes white and flat. Manyray flatfish eat mostly fish and an occasional crab or shrimp.

Yellow Tang *(lau'īpala)*, *Zebrasoma flavescens*. These stunning 8-inch-long surgeonfish are common from shore to beyond scuba limits. Yellow tangs graze on stringy seaweed that grows on hard surfaces. A white spot outlines each of the two scalpel-sharp spines on these yellow fish.

Goldring Surgeonfish *(kole)*, *Ctenochaetus strigosus*. These 7-inch-long fish are found from shallow water to beyond scuba limits. Goldring surgeonfish eat plant and animal material by raking the bottom with their teeth, then sucking the food they find into their mouths. The bright gold ring around each eye makes these brown surgeonfish with blue dots easy to identify.

Brown Surgeonfish *(māʻiʻi),*
Acanthurus nigrofuscus. Brown surgeonfish aren't totally brown: look for
a lavender tinge on the brown body, a
black spot both above and below the
scalpel, and orange dots on head and
chest. These common 8-inch-long surgeonfish are second only to convict
tangs in abundance. Look for brown
surgeonfish in surge areas near the edge
of the reef, where they graze on bottom-growing seaweeds.

Sailfin Tang *(māneoneo), Zebrasoma*
veliferum. Sailfin tangs grow to 15 inches long. They are easily distinguished
by their large back and belly fins, which
the fish open and close while swimming. Look for these striped brown,
orange, and white surgeonfish near
surge areas, where they graze on bottom-growing seaweed.

Whitebar Surgeonfish *(maiko), Acanthurus leucopareius.* These 10-inch-long
surgeonfish graze on seaweed, usually in surge areas. Aptly named, the whitebar
surgeonfish has a brown body marked with a distinct white stripe just behind each
eye and another on the tail just behind each scalpel.

Whitecheek Surgeonfish, *Acanthurus nigricans.* These black fish, which grow to about 8 inches long, have several white markings. Besides white cheek spots, look for a white circle around the mouth and a white tail. Splashes of yellow decorate the fins, the eyes, and the tail spine of this fish. Whitecheek surgeonfish are rare in the Islands but are fairly common in the bay, where they are found in the turbulent water of surge zones. These fish graze on seaweed.

Bluestripe Butterflyfish *(kīkākapu),* *Chaetodon fremblii.* Bluestripe butterflyfish are found only in the Hawaiian Islands, where they are common in the shallow waters of coral reefs, including Hanauma Bay. These butterflyfish eat the tentacles of tubeworms as well as other kinds of invertebrates. Unlike most butterflyfish, this one doesn't have bars running through the eyes. These yellow fish, which are the only local butterflyfish with blue stripes, grow to about 6 inches long.

Threadfin Butterflyfish *(kīkākapu),* *Chaetodon auriga.* This 6-inch-long, yellow-and-white butterflyfish can be easily identified by its trailing fin and black mask. The narrow snout of the threadfin plucks the soft parts of corals, seaweed, and worms from reef cracks.

Raccoon Butterflyfish *(kīkākapu),* *Chaetodon lunula.* These charming fish are common both inside and outside the reef. Look for 8-inch-long, striped butterflyfish wearing black masks outlined in white. Raccoon butterflyfish are most active at night, eating a variety of marine life such as snails, tubeworm tentacles, soft parts of corals, and seaweed. These and most other butterflyfish often swim in pairs.

Stocky Hawkfish *(po'opa'a), Cirrhitus pinnulatus.* Hawkfish are named for their habit of swooping upon passing prey, often crabs. Because the 11-inch-long stocky hawkfish does its hunting mostly at night, look for this fish hiding among coral rocks during the day. This and other hawkfish prop themselves with thickened fins located just behind the gill opening. If you startle them, most hawkfish will glide to another nearby perch, where they resume their quiet surveillance. Stocky hawkfish are white with red spots.

Arc-eye Hawkfish *(piliko'a), Paracirrhites arcatus.* This is one of the few reef fish with markings that seem to emphasize rather than camouflage the eye. Arc-eye hawkfish grow to about 5 inches long and are common from shallow water to beyond scuba limits. Look for these attractive orange-and-brown fish with the orange-and-blue eye markings on top of coral heads, where they feed on crabs that live in the branches.

Blacktail Snapper *(to'au)*, *Lutjanus fulvus*. This 13-inch-long nonnative was introduced to the Hawaiian Islands from the Society Islands in 1956. Although these yellowish gray fish with black tails are established here, they aren't common. Young blacktail snappers are often found in shallow water inside the reef, where they eat crabs and small fish, mostly at night.

Yellowspot Trevally *(ulua)*, *Carangoides orthogrammus*. Adult yellowspot trevally usually live outside the reef, but small schools of young fish sometimes come into waist-deep water. These silvery gamefish with yellow spots occasionally root in the sand for crabs, shrimps, and fish. Yellowspot trevally grow to about 24 inches long.

Bluefin Trevally *('ōmilu)*, *Caranx melampygus*. It's common to see young bluefins inside Hanauma Bay's reef, where they feed on small fish. Bluefin trevally move outside the reef into deeper water as they mature. These fast-swimming silvery predators with blue fins grow to about 36 inches long. Because several kinds of trevally are called *ulua* in Hawai'i, you might hear this fish called a blue *ulua*. Hanauma Bay is one of the few places in the state where you'll commonly see *ulua* swimming among people. In other places, where spearfishing is allowed, these fish are more wary.

TOUR 4: INTERMEDIATE SNORKELING

← Toilet Bowl

Witches' Brew

Telephone Cable Channel

Back Door

FOR

Snorkelers comfortable swimming in mild currents, surge areas, and deep water. You must be able to clear your mask and snorkel while swimming or treading water.

EQUIPMENT

Mask, snorkel, and fins.

DIRECTIONS

Enter the water at the lifeguard station near the center of the sand beach. Swim toward the opening, marked with a "strong current" sign. This is labeled Telephone Cable Channel on the map.

Use the underwater cable that runs through this channel as a guide. Just outside the reef, turn left, then stay in 15 to 20 feet of water as you explore. Watch the shoreline to keep track of where you are. When you're near the end of the sand beach, you have two choices. If the water is calm, you can swim back inside the reef at the channel called Back Door, or you can make a U-turn and return along the reef in 10 to 15 feet of water. If you have any doubts about the conditions, swim back the way you came, then turn right at the cable.

SAFETY TIPS

Make this excursion only during calm or low surf conditions of 1 to 2 feet. If you have doubts about whether you should go, ask the lifeguards. Even when waves are small, currents run seaward from inside the channels. Before going outside the reef, check the strength of the current by turning around in waist-deep water and swimming against it. If you don't make for-

ward progress with ease, don't go out.

If the outgoing current is strong and you have trouble swimming back inside the reef, don't panic. Swim to one side of the channel, staying away from the middle; current is weaker along the edges. Another option is to swim along the outside of the reef to the edge of the bay leading to Witches' Brew, then climb out on the ledge or snorkel in to the beach.

If you feel out of control or see someone else having trouble, shout and wave for help.

Snorkelers will see the greatest diversity of marine life just outside the reef where deep-water and shallow-water environments meet. Here, milletseed butterflyfish, common inside the reef, approach a swimmer in deeper water.

Underwater Telephone Cable. The cable you see on the ocean floor was the first telephone link from Hawai'i to the U.S. mainland. In 1957, AT&T and Hawaiian Telephone laid this California-Hawai'i cable, which is no longer in service.

Cauliflower Coral *(ko'a), Pocillopora meandrina.* It's easy to identify this well-named coral because each colony looks like a head of cauliflower. These pink or tan colonies grow from 10 to 15 inches in diameter, live for about 20 years, then die. Cauliflower coral heads grow in strong surge areas where other corals can't survive. Look in the wavy folds of this coral for crabs, fish, and other marine animals that hide there. An arc-eye hawkfish perches in this particular head.

Coral Reef Diversity. Once outside the reef, notice the different colors of the living corals. The colors come from algae that live within the corals' soft cells. These plants provide some of the food and oxygen that coral animals need to grow; in return, the plants get carbon dioxide and minerals from the corals. In the picture, milletseed and raccoon butterflyfish swim around living coral heads with achilles and convict tangs and brown surgeonfish.

Lobe Coral *(ko'a), Porites lobata.* This is the most common type of coral in Hawai'i. Lobe coral grows in encrusting humps of yellowish green, light brown, or bluish gray. The 6-foot-high mounds so common in the bay are hundreds of years old. Notice the dark cracks that often mark the surfaces of lobe coral colonies (beneath and in front of this white ulua, *Caranx ignobilis*). These cracks are burrows made by snapping shrimps, the animals responsible for the crackling sound you often hear underwater.

Finger Coral *(ko'a), Porites compressa.* The second most common coral in Hawai'i is finger coral. This type varies in shape but usually has projections thicker and longer than human fingers. Finger coral ranges in color from light brown to yellow and grows best in calm, protected water. Sometimes you'll see dead, white base branches of this coral with colored, living tips.

Feather Duster Worm, *Sabellastarte sanctijosephi.* The picture shows the center of a feather duster worm's fan, the creature's food-filtering and oxygen-gathering organ. These "feather dusters" can grow to about 6 inches wide in a multitude of colors and patterns. When disturbed, the worm withdraws its pretty fan into a tube, where the body of the worm always stays hidden. Feather duster worms build permanent tube homes inside rock cracks, usually in areas with sediment in the water. Look for feather dusters in places where silty deposits cover rocks.

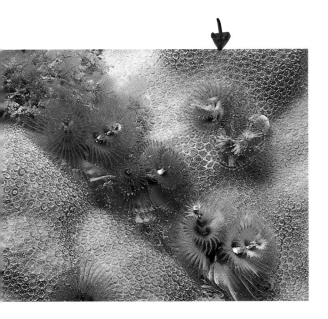

Christmas Tree Worm, *Spirobranchus giganteus.* Young Christmas tree worms settle onto living coral, then build hard-shelled tubes as the coral grows around them. The colorful spiral fans that poke out of the tubes filter tiny, drifting plants and animals from the water. These "trees" can be yellow, blue, or reddish orange. When startled, the worms withdraw their half-inch long fans into their tube homes.

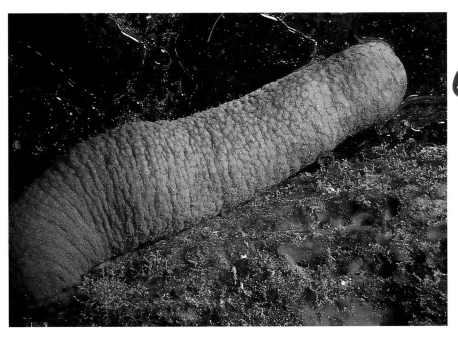

Brown Sea Cucumber *(loli), Actinopyga obesa.* Although they don't look it, sea cucumbers are close relatives of sea urchins and sea stars (starfish). Sea cucumbers clean sand by stuffing it into their mouths with tentacles, digesting the decomposing plant and animal material in it, then excreting the sand in lines that look like strands of sandy pearls. If you see these bumpy lines in the sand, you know a sea cucumber has been grazing there. The brown sea cucumber grows to about 12 inches long.

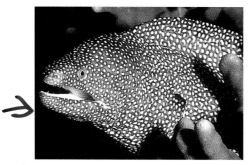

Whitemouth Moray Eel *(puhi 'ōni'o), Gymnothorax meleagris.* Look in coral cracks and crevices for moray eels. It's common to see morays backed into reef holes with heads up, swaying back and forth. Hawai'i hosts more than 35 kinds of moray eels; many of these live among the reefs of Hanauma Bay. The whitemouth moray, which grows to 3-1/2 feet long, eats mostly fish. A brown eel with white spots, the white-mouth is one of the most common moray eels in Hawai'i.

Yellowmargin Moray Eel *(puhi paka), Gymnothorax flavimarginatus.* Another common eel in Hawai'i is the yellowmargin moray eel, which grows to about 4 feet long. These eels, speck-led brown with yellowish edges to their fins, eat fish and occasionally crabs. The continual jaw movement of these and other eels give them a menacing appearance, but this is simply the way these fish breathe; they're pumping water into the mouth and out the gills.

Stout Moray Eel *(puhi), Gymno-thorax eurostus.* These common brown eels with various-colored spots live among the rocks and coral of the reef from 5 to 50 feet deep. Stout morays, which grow to about 2 feet long, mostly eat small fish. Although moray eels aren't usually aggressive toward humans, they may bite if you put a hand or foot near their mouths.

Zebra Moray Eel *(puhi), Gymno-muraena zebra.* Not all moray eels have sharp teeth. These zebra morays have pebblelike teeth that effectively crush crabs and other shelled animals. Zebra morays, which are brown with yellow stripes, can grow to nearly 5 feet long.

Trumpetfish *(nūnū), Aulostomus chinensis.* These narrow fish grow to at least 27 inches long, but some in the bay may seem even longer. Both gray and yellow trumpetfish are different color phases of the same species. Look for trumpetfish lying motionless in the water, sometimes floating upright like sticks. These fish eat shrimps, crabs, and small fish.

Trumpetfish Eating. When an unsuspecting fish, like this squirrelfish, swims by, the trumpetfish sucks in, vacuuming up the prey with its long tube mouth. While waiting for this shot, the photographer watched this trumpetfish slowly change from bright yellow to this grayish color.

Cornetfish, *Fistularia commersoni.* You can easily tell a cornetfish from its trumpetfish relative: A cornetfish has a trailing filament from its tail, but a trumpetfish does not. Cornetfish are greenish in color and can grow to a startling length of 5 feet or more. However, most cornetfish in the bay are usually closer to 2 or 3 feet long. Cornetfish eat crabs, shrimps, and small fish.

Yellowfin Goatfish *(weke 'ula), Mulloides vanicolensis.* These 15-inch-long goatfish are pinkish with yellow fins and are usually found in deeper water than their yellowstripe relatives. Yellowfin goatfish often feed in groups during the day but individually at night, eating crabs, shrimps, fish, worms, and other invertebrate animals. To identify any goatfish, look for the two whiskerlike barbels on the chin that the fish use to sense food.

Whitesaddle Goatfish *(kūmū), Parupeneus porphyreus.* The white saddle in this goatfish's name refers to the light spot just in front of the tail. These 15-inch-long reddish fish, which are found only in Hawaiian waters, probe rocks, reef, and sand with their barbels to find mostly shrimps and crabs.

Bluespine Unicornfish *(kala), Naso unicornis.* All unicornfish are surgeonfish, but not all unicornfish have a "horn." The 27-inch-long bluespine unicornfish, however, is true to its name, sporting one hornlike projection from its forehead. Researchers don't know how, or if, the fish uses this projection. Bluespine unicornfish, which are gray with blue trim on their fins and around their "scalpels," eat leafy seaweeds.

Whitespotted Surgeonfish *('api) Acanthurus guttatus.* These striking 11-inch-long fish are found in areas of strong surge and breaking waves. Here, small schools of whitespotted surgeonfish graze on seaweed. It's possible that the white spots on these brown surgeonfish resemble the bubbles of their environment and thus help the fish blend into their background.

Orangeband Surgeonfish *(na'ena'e), Acanthurus olivaceus.* Look for these distinctive surgeonfish cruising over sandy bottoms near the reef. Orangeband surgeonfish, which grow to 12 inches long, eat diatoms, seaweed, and decaying plants and animals. Like some surgeonfish relatives, these fish also swallow some sand with their meals, which may help digestion. Look for brownish fish with a bright orange band just behind each eye.

Forcepsfish *(lauwiliwili nukunuku 'oi'oi), Forcipiger flavissimus.* When startled, this 7-inch long, black-and-yellow butterflyfish often swims upside down and backwards. The long snout of this fish enables it to reach into reef cracks to eat worms, shrimps, the tube feet of sea urchins, and fish eggs.

Teardrop Butterflyfish *(lauhau), Chaetodon unimaculatus.* You can identify this 7- to 8-inch-long butterflyfish by the black, teardrop-shaped spot on each side of its light yellow body. Like many of its butterflyfish cousins, teardrop butterflyfish often swim in pairs, eating coral, worms, other invertebrates, and seaweed.

Oval Butterflyfish *(kapuhili), Chaetodon trifasciatus.* Like many other butterflyfish, this type usually swims in pairs near coral reefs, grazing on the soft bodies of coral animals. Oval butterflyfish, which have dark horizontal stripes on orange bodies, grow to 6 or 7 inches long. To identify the many butterflyfish of the bay, look for spots on the body and remember if stripes are horizontal, vertical, or diagonal.

Fourspot Butterflyfish *(lauhau), Chaetodon quadrimaculatus.* These butterflyfish have two white spots on each side of their brown-and-orange bodies, hence their common and scientific names. It's common to see these 6-inch-long butterflyfish swimming in pairs along the reef and near the bottom. They commonly eat the soft bodies of corals.

Saddleback Butterflyfish *(kīkākapu),*
Chaetodon ephippium. These stunning
8-inch-long fish aren't common in
Hawai'i, but you can sometimes see
them grazing on coral and seaweed just
outside Hanauma's reef. Saddleback butterflyfish are easy to spot: look for a
white fish with a black "saddle" on its
back. The trailing back fin adds to this
fish's graceful beauty.

Multiband Butterflyfish *(kīkākapu),*
Chaetodon multicinctus. Multiband
butterflyfish are found only in Hawai'i.
These 4-inch-long fish commonly eat
the soft bodies of stony corals. Look for
small, off-white fish decorated with
brown bars and brown spots; these fish
are only tinged with yellow.

Ornate Butterflyfish
(kīkākapu), Chaetodon
ornatissimus. It's easy to
spot these striking butterflyfish when they're in the
vicinity: they look like
cheerful, painted clowns.
These 7-inch-long fish,
which often swim in
pairs, are white with diagonal orange stripes. They
eat mostly the soft bodies
of corals.

Yellowtail Coris *(hīnālea 'akilolo), Coris gaimard.* Female. The colors and markings on male and female yellowtail coris vary somewhat, but both have bright blue spots and yellow tails. These lovely members of the wrasse family are usually found near the reef eating snails, crabs, and hermit crabs. Yellowtail coris grow to about 15 inches long.

Yellowtail Coris *(hīnālea 'akilolo), Coris gaimard.* Juvenile. This young fish looks so different from its adult phases that researchers once thought it was a separate species. For a time, adolescent yellowtail coris wear the colors of both juveniles and adult females, an unusual sight to behold. You can see that this red juvenile is just beginning its change to adult colors.

Elegant Coris, *Coris venusta.* This wrasse, found only in Hawai'i, is aptly named. Look for these red-and-yellow, 7-inch-long fish in the more shallow areas outside the reef. Here they search for snails, crabs, sea urchins, shrimps, worms, and other invertebrates.

Spotted Pufferfish *('o'opu hue), Arothron meleagris.* Pufferfish get their name from their ability to suck water or air into their bodies and puff themselves up. This inflated shape can be hard for predators to swallow. Spotted pufferfish, which are brown with white spots, can grow to about 13 inches long. These fish eat mainly coral but will also eat invertebrates, seaweed, and decomposing plant and animal material.

The skin, organs, and sometimes the flesh of pufferfish can contain a deadly poison that can kill people even if they eat only a small amount. In Japan, some consider pufferfish, called *fugu*, a delicacy, but eating it is risky. For a few unlucky diners, *fugu* is their last meal.

Spiny Pufferfish *('o'opu 'ōkala), Diodon holocanthus.* Spiny pufferfish belong to the porcupinefish family, so-called because of the spines that cover their heads and bodies. When this fish inflates itself, the spines stand up, a hindrance to most predators. Spiny pufferfish, which grow to about 15 inches long, are light brown with dark brown spots. These fish have strong jaws that can crush the shells of snails, sea urchins, and crabs.

49

Porcupinefish *(kōkala), Diodon hystrix.* The picture shows how these fish got their common name. Porcupinefish, which can grow to 28 inches long, usually hunt at night, hiding in holes and under ledges during the day. Like their pufferfish relatives, the skin, organs, and sometimes the flesh of porcupinefish can contain a deadly toxin. The diet of these light brown fish with black spots includes snails, crabs, and sea urchins.

Reef Triggerfish *(humuhumu-nukunuku-ā-pua'a), Rhinecanthus rectangulus.* Triggerfish have a long, movable bone on the belly and three spines on the back. The forward spine can be locked in an upright position by the second spine, the "trigger." When danger threatens, the fish backs into a hole in the reef and wedges itself in, erecting the spines and pelvic bone. In such a position, it's nearly impossible to pull the triggerfish out. The long Hawaiian name of this 10-inch-long fish means "fish with a snout like a pig." Reef triggerfish, which are mostly brown, black, and blue, grunt like pigs when cornered or caught. These fish eat nearly anything that they find on the ocean floor, including seaweed, decomposing plants and animals, and invertebrates.

Lei Triggerfish *(humuhumu lei)*, *Sufflamen bursa.* This tan, 8-inch-long fish gets its common name from the brown bars that run through and just behind the eyes, resembling a lei around the fish's "neck." Lei triggerfish are found at depths from 10 feet to beyond scuba limits, searching the bottom for just about anything edible they can find. You'll see solitary lei triggerfish swimming just outside the reef, but it might be hard to get a close look at them. These wary fish tend to take off and hide when swimmers approach.

Pinktail Durgon *(humuhumu hi'u kole)*, *Melichthys vidua.* Durgons are members of the triggerfish family, having similar back-and-lock systems. These 13-inch-long pinktail durgons are easy to spot. The dark brown fish, tinged with yellow, have lovely white rear fins and pinkish white tails. Durgons usually eat seaweed and decomposing plants and animals but also feed on living fish and invertebrates.

Black Durgon *(humuhumu 'ele'ele)*, *Melichthys niger.* A close look at these 12- to 13-inch-long fish shows that they aren't really black, but a deep, blue green. Light blue bands lining the rear fins complete the unique color scheme of these striking fish. Like their triggerfish relatives, black durgons hide and lock in place when threatened. However, because of their protected status in Hanauma Bay, some of these triggerfish are not as wary of humans as they are in unprotected places. Hanauma Bay is an ideal place to get a good look at these and other members of the triggerfish family, which eat mostly seaweed.

Fantail Filefish *('ō'ili 'uwī'uwī), Pervagor spilosoma.* You can tell filefish from triggerfish by the location of the first dorsal spine: in filefish it's directly above the eye; in triggerfish it's well behind the eye. Fantail filefish have locking-spine systems similar to those of their triggerfish cousins. Fantail filefish grow to 7 inches long and are found only in Hawai'i. Their yellow bodies are covered with black dots. Every few years, the fantail filefish population explodes, and, for a while, these little fish seem to be everywhere. Then, presumably because their numbers outgrow their food supply (seaweed, decomposing tissue, and invertebrates), the fish die and wash ashore in great numbers.

Barred Filefish *('ō'ili),*
Cantherhines dumerilii.
These 14-inch-long, gray-brown fish are easy to identify by their mouths: white lips lined in black outline large, protruding teeth. Barred filefish use these remarkable teeth to eat branching-type corals, but they also munch on bottom-dwelling animals. Here, a yellow trumpetfish hovers just above a barred filefish.

Agile Chromis, *Chromis agilis.* These 4-inch-long damselfish are found at depths from 15 feet to beyond scuba limits. To identify these common fish, look for a brown body tinged in pink with a dark spot at the base of each side, or pectoral, fin. These and other chromis eat tiny animals adrift in the ocean.

Blue-eye Damselfish, *Plectroglyphidodon johnstonianus.* Blue-eye damselfish are true to their name, with light blue markings around and in front of the eyes. You can also identify these 4-inch-long fish by blue fin outlines and a large dark spot on each side of the yellowish body, just in front of the tail. Blue-eye damselfish mostly eat the soft bodies of live coral.

Halfbeak, Hemirhamphidae Family. Halfbeaks are so called because their upper jaws are shorter than their lower jaws, which are long and flat. These silvery surface fish eat seaweed, fish, and animal life adrift in the ocean. Because halfbeaks, which are related to needlefish and flyingfish, are found near the water's surface, it's easy to miss them while snorkeling. To see halfbeaks and needlefish, remember to look up occasionally toward the surface. Halfbeaks grow from 10 to 20 inches long.

TOUR 5: ADVANCED SNORKELING

Toilet Bowl

Witches' Brew

Telephone Cable Channel

Back Door

FOR

Strong swimmers comfortable with free diving. Must be able to swim long distances in deep water, with swells and against currents. This can also be a shallow scuba dive.

EQUIPMENT

Mask, snorkel, and fins.

DIRECTIONS

Swim out the channel marked with a "strong current" sign and labeled Telephone Cable Channel on the map. Follow the right side of the cable to just

inside Witches' Brew, then loop back, swimming along the wall. Go only as far as you feel comfortable. Don't swim farther than Witches' Brew; the water past there is too deep for viewing while snorkeling. Because some of the animals that live in this zone are bottom dwellers or hide under ledges, you'll need to free dive to see them. When you spot something interesting, get a fix on where it is, then dive to it.

SAFETY TIPS

Check with lifeguards before taking this

tour. You should go only during calm or small surf conditions of 1 to 2 feet. Not only can bigger waves be dangerous, but stirred up sediment clouds the water and spoils the view.

Watch the land to monitor how far out you are swimming. Be aware of how tired you feel. Remember that at your farthest point from land, your swim is only half over. Save enough energy to get back.

Always make this trip with a strong swimming partner who is also comfortable with swells, currents, and deep water. If you get tired, take a rest by floating motionless either on your back or face down while breathing through your snorkel.

If you feel uncomfortable in the deep water, tell your partner, then both swim calmly toward the closest edge of the bay. Wave and shout if you need help.

WHAT TO LOOK FOR

Antler Coral, *Pocillopora eydouxi.* This brown coral grows at 20 feet and deeper in clumps up to 4 feet high. Its branches look like moose antlers. Look for schools of damselfish, like these Hawaiian *Dascyllus* hovering around antler coral. When startled, the fish dive into the shelter of the outstretched branches.

Antler Coral Close-up. Antler and all other stony corals consist of soft sacklike bodies surrounded at the top by stinging tentacles. Each body sits in a cup of hard limestone, the animal's skeleton. Corals grow into colonies when individuals pinch off each other to make new bodies, then the skeletons fuse. Because a thin layer of tissue grows over the entire skeleton, connecting all the polyps in a colony, it's important to keep hands and feet off all coral. Touching it may damage or kill it.

Slate Pencil Urchin *(pūnohu)*, *Heterocentrotus mammillatus.* This bright red sea urchin is found in both exposed and protected places on the reef. The animal gets its common name from the color that readily rubs off its spines. (Slate pencil was once another name for chalk.) At the center of the underside, each urchin has a mouth with five tiny teeth. Slate pencil urchins graze mostly on bottom-growing seaweed. Their spines grow to about 4 inches long.

Collector Urchin *(hāwaʻe)*, *Trip-neustes gratilla.* These black, short-spined urchins get their common name from their practice of piling everything from trash to seaweed on their shells, then holding the collection there with tube feet. Researchers believe that this behavior is the result of the animal's aversion to bright light. Because of this light sensitivity, it's common to see collector urchins nestled in the shade of reef overhangs. These seaweed grazers can grow to about 6 inches across.

Spiny Urchin *(wana)*, *Echinothrix diadema.* Spiny urchins have two kinds of spines. The longest are sharp and hollow, and the shorter are barbed and venom-tipped. Because these spines are movable, the animals can tilt and wave them in the direction of intruders. The spines of these seaweed grazers serve as defense. Spiny urchins grow to at least 10 inches in diameter. This type is black or dark purple.

Cushion Sea Star, *Culcita novae-guinae.* Sea stars (starfish) and sea urchins are close relatives. All use tube feet to move about and attach themselves to surfaces. The multicolored cushion stars grow to about 10 inches across and graze on the soft bodies of corals. Because these plump, pillowlike creatures eat only small amounts of coral at one time, the coral colony is usually able to recover from a cushion star attack without permanent damage.

Crown-of-thorns Sea Star, *Acanthaster planci.* These sea stars, which grow to at least 12 inches across, are notorious in some parts of the Pacific for eating coral in such large amounts that the reefs suffer extensive damage. So far, Hawai'i's crown-of-thorns population is small enough that the creatures don't threaten local reefs. This sea star's thorns, which carry a mild toxin, can puncture skin. Crown-of-thorns sea stars are green or dull red.

Comet Sea Star, *Linckia multifora.* Comet sea stars get their name from the fact that they can grow an entire body from one arm or even a piece of one arm. Because the original arm is longer than the new ones during this growth period, the animal resembles a comet. Since these animals often reproduce this way, few comets are ever symmetrical. Mature comet sea stars are about 4 inches across.

Green Sea Star, *Linckia diplax*. This sea star grows to about 12 inches from arm tip to arm tip. Green sea stars are close relatives of comet sea stars but are larger and live in deeper water. You can identify green sea stars and their cousins by their smooth, leathery upper surface and long, thin arms.

Orange Sea Star, *Mithrodia fisheri*. This sea star is less common than comet sea stars and usually lives in deeper water. Because most sea stars turn their stomachs inside out to digest prey externally, picking them up to look at them can interrupt the creature's meal. It's best to enjoy these symbols of the sea without disturbing them. Orange sea stars grow from 6 to 10 inches across.

Textile Cone Shell *(pūpū 'alā), Conus textile.* Cone shell snails are predators that eat worms, fish, or other snails. Most cone shells, like this textile cone, inject a potent toxin through a hollow tubelike structure that paralyzes or slows down their prey. The snail then expands its mouth around the subdued prey and digests it. The sting of these snails can wound people, sometimes severely. Even though these brown-and-white-patterned shells look beautiful, resist the temptation to handle any cone-shaped shell. Look for cone shells on the ocean floor or resting on a coral head like this one. Sizes vary from less than 1 inch to about 3 inches long.

Jellyfish, *Cephea cephea.* All true jellyfish swim by contracting their bells; their trailing tentacles sting and collect fish or animals adrift in ocean currents. Some but not all jellyfish tentacles can sting people. Because reactions to these stings vary from person to person, it's safest to look at but not touch jellyfish. No one knows the function of the unusual projections at the top of this bluish 4- to 6-inch-wide jellyfish, which has no common name. These and other jellyfish occasionally drift into the bay but are rare.

Hermit Crab *(unauna), Dardanus sanguinocarpus.* You can easily distinguish between living snails and hermit crabs in snail shells: the crabs walk much faster than snails ever could. This pink, red, and white hermit crab has adopted an empty cone shell for its home. Most hermit crabs are scavengers, eating decomposing plant and animal tissue.

Hairy Hermit Crab *(unauna), Aniculus maximus.* Hermit crabs are experts at recycling. When a hermit crab outgrows its adopted shell home, the creature upgrades to a larger one. Because empty shells are essential to these animals and therefore to the health of the reef, it is illegal to collect shells (or anything else) at Hanauma Bay. The largest hairy hermit crabs, which have hairy, yellow legs, live in triton's trumpet shells, which grow up to 20 inches long.

Swimming Crab *('ala 'eke), Charybdis hawaiiensis.* This active crab has two paddle-shaped legs at the rear that propel the animal around the reef. But watch out for the front of this crab: if it is cornered, its sharp fish-catching pincers can grab fingers or anything else that comes within reach. Look for this common, 3-inch-wide brown crab hiding in the folds of branching corals or in the sand.

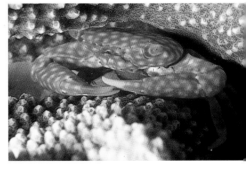

Coral Crab, *Trapezia* sp. These red crabs with white spots also live between coral branches. If you see two coral crabs, they are most likely a male and a female, their common living arrangement. Coral crabs eat tiny fat particles produced by the coral. In return, the crabs protect the coral by fending off hungry sea stars. Coral crabs are usually from 1 to 2 inches wide.

Spiny Lobster *(ula), Panulirus marginatus.* These dark lobsters are social animals that gather under rocks and ledges, where they pass the day together. At night, these predators venture out, searching the reef for snails, clams, crabs, and shrimps. Spiny lobsters have long, stout antennae that often poke out from the edge of their daytime hiding places. These black, waving antennae are sometimes all you will see of these lobsters, which can grow to 18 inches long.

Regal Slipper Lobster *(ula papapa), Arctides regalis.* Slipper lobsters have wide, flat bodies with mottled colors that blend with their background. Of the eight types of slipper lobsters found in Hawai'i, the 6-inch-long regal is the most colorful. Like the spiny lobsters, all slipper lobsters are nocturnal, venturing from their daytime resting places to hunt snails, clams, shrimps, and crabs. Neither spiny nor slipper lobsters have enlarged pincers, as do their New England cousins, the American lobsters.

Hawksbill Sea Turtle *('ea), Eretmochelys imbricata.* Compared with green sea turtles, hawksbills are rare in Hanauma and in Hawai'i. The few hawksbills that do live here are found close to coral reefs, where they poke their narrow beaks into crevices for sponges and other invertebrates. Hawksbill sea turtles are smaller and have more elongated beaks than green sea turtles. Over the years, people have slaughtered hawksbill turtles to near extinction for their lovely shells. If you are lucky enough to see one, remember that this turtle is rare and needs our protection. Don't chase; just watch. Adult hawksbill shells, "tortoise shell" in color, grow to about 3 feet long.

61

Green Sea Turtle *(honu), Chelonia mydas.* Of the several types of sea turtles found in Hawaiian waters, only the green sea turtle is common in the bay. These turtles are named for their green body fat once used in soup. Their shells are gold, brown, and black. Look for green sea turtles floating on the water's surface, resting underwater, or grazing on submerged seaweed. Because animals have been protected in the bay since 1967, and turtles are protected throughout the state, some turtles now commonly approach swimmers. Enjoy the beauty and charm of these creatures without touching, chasing, or riding them. This is not only illegal but could harm the animal. Green sea turtles grow 3 to 4 feet long and can weigh up to 400 pounds.

Stingray *(hīhīmanu)*, *Dasyatis* sp. Stingrays are disc-shaped, brownish fish that cruise the bottom of the bay in search of snails and other bottom dwellers. This common stingray can reach a width of 3 feet, but most are smaller. When not feeding, stingrays lie motionless, often covered with sand. It's during these resting times that people accidentally step on them, sometimes resulting in a painful experience since stingrays have a venomous spine on the upper part of the tail. These and other rays only use these spines defensively and never pursue humans. Rays are closely related to sharks.

Spotted Eagle Ray, *Aetobatus narinari.* Other common rays in the bay are eagle rays, which look like underwater kites. The eagle ray's upper, slate gray surface is covered with white spots; the underside is white. Eagle rays measure 2 to 4 feet in width, each bearing a venomous spine at the base of a long tail. As in stingrays, this spine is for defense only. Eagle rays, with duck-billed snouts cruise over sandy areas of the bay digging up shelled animals and sea urchins

Glasseye *(ʻāweoweo)*, *Heteropriacanthus cruentatus.* Glasseyes, which grow to about 12 inches long, are named for their large, transparent eyes. These fish are nocturnal, hovering in dark recesses of the reef during the day, then hunting for small fish and invertebrates at night. Look under ledges during the day for these red fish.

Evermann's Cardinalfish *('upāpalu),*
Apogon evermanni. Cardinalfish get
their name from the red members of
the family even though several are gray
and tan. Most cardinalfish are nocturnal, hiding in caves during the day, then
foraging for tiny, drifting animals at
night. Peek into reef caves during free
dives to glimpse cardinalfish. Hawai'i's
cardinalfish range from 2 to 7 inches
long and live at depths from 3 feet to
beyond scuba limits. Male cardinalfish
incubate the eggs in their mouths.

Shoulderbar Soldierfish *('ū'ū),*
Myripristis kuntee. Soldierfish, members of the squirrelfish family, are often
known in Hawai'i by their Japanese
name, *menpachi.* Like other squirrelfish, soldierfish are red and nocturnal, hiding in dark places during the
day, then foraging for crabs and shrimps
at night. Soldierfish have a streak of
dark pigment just on or behind each gill
opening. Of the three common soldierfish found in Hawaiian waters, the
shoulderbar soldierfish is the smallest,
growing to just over 7 inches long.

Sleek Unicornfish *(kala holo), Naso*
hexacanthus. These unicornfish have
no "horn" on their foreheads but resemble other unicornfish in body and fin
shapes. Sleek unicornfish can grow to
about 30 inches long but are usually
smaller. They are most common in
water 50 feet and deeper. In Hanauma
Bay, they are sometimes found in shallower water along steep reef slopes.
Look for these fish in several colors,
ranging from dark brown to pale blue.
Sleek unicornfish eat tiny animals adrift
in ocean currents.

Crown Toby, *Canthigaster coronata.* These pufferfish grow to about 5 inches long. People usually see crown tobies at 75 feet or deeper, but in the bay, crown tobies are often found in more shallow water closer to shore. Look for a slow-swimming white fish with yellow spots, black bars, and a pointed snout. Crown tobies eat invertebrates and seaweed from the ocean floor.

Pennantfish, *Heniochus diphreutes.* These 8-inch-long butterflyfish are often found in water over 50 feet deep, but because they stay well above the bottom, they're usually easy to spot while snorkeling. Look for these white-bodied, black-banded pennantfish at the outer reef off exposed coasts of the bay, where they feed on tiny, drifting animals.

Pyramid Butterflyfish, *Hemitaurichthys polylepis.* Look for these 6-inch-long, yellow-and-white butterflyfish near their pennantfish cousins. Pyramid butterflyfish are often found in groups at 50 feet or deeper but hang in the middle of the water column as if suspended. From a distance, the white bodies of these fish stand out like glowing triangles. Pyramid butterflyfish eat tiny animals that drift in ocean currents.

Variegated Lizardfish *('ulae), Synodus variegatus.* At 10 inches long, these lizardfish are slightly smaller than the orangemouth lizardfish but are more common. Look carefully. Even when variegated lizardfish aren't hiding in the sand, they blend well with their background.

Variegated Lizardfish Buried in Sand. After being startled by the photographer, the above lizardfish dived into the sand, then lay motionless.

Orangemouth Lizardfish (*ʻulae*), *Saurida flamma*. Lizardfish can be hard to find because they blend with the ocean floor, sometimes even burying themselves in sand. Like their namesake, these fish wait motionless for passing prey, usually small fish, then strike. Orangemouth lizardfish are the largest lizardfish you will see in the bay, growing to about 12 inches long.

Reticulated Butterflyfish, *Chaetodon reticulatus*. If you get a glimpse of this black, gray, and yellow beauty, it's your lucky day. These 7-inch-long fish are rare in Hawaiʻi, but some do live in Hanauma Bay. Look for these butterflyfish with the reticulated, or lacelike, pattern on their bodies eating coral or grazing on seaweed.

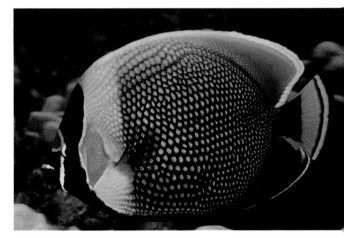

Amberjack (*kāhala*), *Seriola dumerilii*. These popular gamefish are usually found far offshore in deep water, but occasionally they come into the shallow waters of Hanauma Bay. Look for sleek, fast-swimming, silvery fish with deeply forked tails and one dark stripe running through each eye. Amberjacks, which grow to at least 4 feet long, feed on fish.

TOUR 6: SCUBA DIVING

Toilet Bowl

Most Difficult

Witches' Brew

Easy

Intermediate

Back Door

Telephone Cable Channel

FOR

Certified scuba divers.

EQUIPMENT

Scuba gear and dive booties that fit inside fins. Divers with low cold tolerance should wear either a full or top half wet suit depending upon preference.

DIRECTIONS

If walking around in scuba gear isn't for you, take the shuttle bus both up and down the hill. There's a minimal charge for this service.

Although you can have a rewarding dive nearly anywhere in the bay, this chapter deals with several areas too deep for good snorkeling, from 30 to about 80 feet deep. Divers have three possible entry/exit points. Choose the one that's right for you:

1. Easy. Snorkel straight out Telephone Cable Channel, following the cable. Dive at about 30 feet, just past the Witches' Brew point. Circle left and explore the middle of the bay to about 50 feet. Head back inside the reef using the cable as a guide.

2. Intermediate. With your gear on, turn right at the beach, then walk the ledge toward Witches' Brew. Just before you get to the point, look for the area of flat, relatively protected water at the edge. Enter here. Snorkel away from the ledge, then out and around the point. Descend there in about 30 feet of water, explore in a circle to about 60 feet or until you've used about half your air, then head back. Exit at the entry point. If you prefer, you can snorkel in through the Telephone Cable Channel by following the cable, but this can be a long, tiring trip.

3. Most Difficult. This is for experienced divers only; the walk is long and currents can be strong, often pushing you offshore. Walk in gear to the Toilet Bowl area, then enter the water along the inlet that leads to the bowl. Snorkel out the inlet, turn left along the shelf, then descend in about 50 feet of water. Explore the outer edge of the reef, then circle back, exiting where you entered.

SAFETY TIPS

Dive in the bay only during calm or small surf conditions of 1 to 2 feet. Big surf creates dangerous currents, waves

A diver explores some geological formations called potholes, formed when a piece of rock rotates during storms and grinds away the rock floor. These potholes are located near the wall just outside Witches' Brew.

on ledges, and cloudy water. Check with the lifeguards if you have any doubts about when or where to dive.

If you feel a current after you drop to the bottom, swim against it for the first part of the dive. Later, you can drift with the current back to your starting place.

Always follow the dive tables or your dive computer and always dive with a partner. If you or your partner can't get back to an edge of the bay, inflate your buoyancy compensator, then wave and shout for help. Above all, stay calm.

A green sea turtle tags along with a diver in Hanauma Bay. Remember to enjoy the company of Hanauma Bay's turtles without touching or chasing them. Foreground: blue-eye damselfish in antler coral.

WHAT TO LOOK FOR

Orange Tube Coral, *Tubastraea coccinea.* This type of coral, which looks like bouquets of orange flowers, is not in the reef-building business. Orange tube corals live in clumps at different depths, from shady tide pools to deep caves. A typical clump contains ten to twenty individuals and is 2 to 4 inches across. Because tube corals grow on dark surfaces where other corals can't compete and sediment can't cover them, you'll most often see these orange animals on cave ceilings and on steep, deep walls. Tube corals eat tiny drifting animals that the corals sting with tentacles encircling their mouth.

Octocoral, *Sinularia abrupta.* Like their coral relatives, octocorals come in a wide range of shapes, sizes, textures, and colors. All, however, have eight tentacles around their mouths that sting tiny, drifting animals, and eight membranes that divide their bodies into compartments. This particular type of octocoral grows from about 30 feet deep to well beyond scuba limits. The texture of the bluish colony looks hard but is actually spongy to the touch.

Flatworm, *Pseudoceros* sp. Flatworms can be brightly colored, like this one, or camouflaged to match their background. You can tell a flatworm from a nudibranch (sea slug) by the thickness of the animal: the worms are thinner and flatter than the nudibranchs. Look for this 1-inch-long flatworm under rocks and ledges in 30 to 60 feet of water. Flatworms eat invertebrates like corals, snails, crabs, shrimps, and other worms.

Tiger Cowrie *(leho kiko), Cypraea tigris.* This 7-inch-long, black-and-brown-spotted cowrie is Hawai'i's largest and one of the most common, living at depths of 12 to 120 feet. About thirty-four different kinds of cowries are found in Hawai'i. Female cowries protect their eggs, covering them with their muscular walking organ, or foot, for 1 to 2-1/2 weeks until they hatch. Look for cowries in reef holes and under ledges. You may see these snails partly covered by their mantle, as shown here. Besides helping camouflage the animal, this thin membrane builds up and polishes the creature's shell. (Remember it is illegal to collect any shells, dead or alive, in this preserve.)

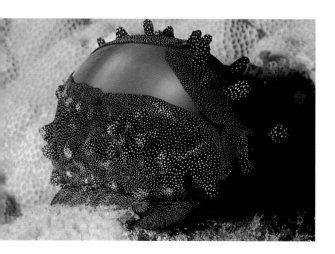

Mole Cowrie *(leho), Cypraea talpa.* This brilliant gold cowrie is not rare but is often hard to find during the day. Mole and other cowries are nocturnal, hiding in the daytime with their camouflaged mantles drawn over the shells, then wandering around at night eating seaweed, sponges, or both. Mole cowries, which live under rocks in 15 to 30 feet of water, grow to just under 3 inches long.

Spanish Dancer Nudibranch, *Hexabranchus sanguineus.* Nudibranchs are sea slugs, snails without shells. This red nudibranch, which can grow to 7 inches long or more, eats sponges. The frills near the rear of the body are the creature's gills. Spanish dancers get their common name from the way they swim, bending back and forth in graceful crescent shapes while flaring out their rolled-up edges. The pink flowerlike object in the picture is this nudibranch's egg mass.

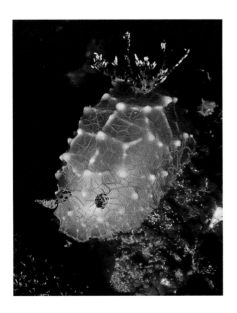

Goldlace Nudibranch, *Halgerda terramtuentis.* This yellow-and-white dorid nudibranch carries its lacy gills on its back. Goldlace nudibranchs, which grow to about 1-1/4 inch long, live at depths from 6 to 90 feet, where they eat sponges.

Eolid Nudibranch, *Pteraeolidia ianthina.* Eolids are a group of nudibranchs with fingerlike projections along their backs. These fingers store stinging cells ingested when the nudibranch eats hydroids, relatives of the Portuguese man-of-war. Algae in the projections can cause color differences in some individuals. This lovely lavender eolid nudibranch grows to about 4 inches.

Dorid Nudibranch, *Phyllidia varicosa.* The word dorid refers to a large group of nudibranchs, most of which have a circle of gills on their backs. This particular dorid is an exception; its gills are under its edges, along the sides of the animal. This common black-and-blue nudibranch, which can grow to over 3 inches long, eats sponges that contain poison. The nudibranch stores the strong-smelling poison for its own defense.

Night Octopus *(he'e pūloa), Octopus ornatus.* Fishermen in Hawai'i often call octopuses squid, although the two are different animals. Two types of octopuses are common on Hawai'i's reefs: a gray type called the day squid *(he'e mauli),* and this orange-brown type called the night squid *(he'e pūloa).* These common English names describe the animals' preferred hunting time. Octopuses eat snails, crabs, shrimps, and fish by enclosing them in their arms, paralyzing them with a salivary toxin delivered by a strong beak, then carrying the prey to their den to eat.

Although they can grow larger, adult day octopuses average about 2 feet long from the top of the head to the bottom of outstretched arms. The night octopus is usually somewhat smaller and thinner.

Octopus Eye. Octopuses often hide in reef holes, poking an eye out at curious divers. Studies show that among all the invertebrates, octopuses are the most apt pupils, learning by sight as well as by touch.

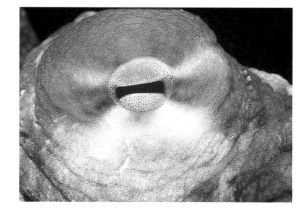

Sponge Crab *(makua-o-ka-līpoa),*
Dromidiopsis dormia. It's easy to pass
near a sponge crab without seeing it.
These deception specialists hold a
sponge, or sometimes a piece of trash,
on their backs with the rear two pairs
of legs. When the crab molts, it sets its
sponge aside, then picks it up again
later to carry on its new shell. The
sponge helps the dark, furry crab blend
in with its reef background. Sponge
crabs can grow to about 10 inches
across.

7-11 Crab *('alakuma), Carpilius mac-*
ulatus. Eleven reddish brown spots
decorate the shell of this crab, which
grows to about 6 inches across the
shell. Look for the 7-11 crab wedged in
holes in the reef, but don't get your fingers near its claws. The creature uses
these strong pincers to break open snail
shells.

Harlequin Shrimp,
Hymenocera picta. If you
look carefully into nooks
and crannies of the reef,
you might be lucky enough
to find these white, red-
spotted, 1-inch shrimps.
Harlequin shrimps usually
live and feed in pairs,
with sea stars being their
favorite meal. These beauties are now scarce in
Hawai'i because they've
been collected extensively
for aquariums. Because of
the park's protected status,
Hanauma Bay is a likely
place to see these shrimps.

Barber Pole Shrimp, *Stenopus hispidus.* Another shrimp that is often found in pairs is the red-and-white-striped barber pole shrimp. These 2-inch-long shrimps range from wading depths to beyond scuba limits, but, because they often hide, you'll see these shrimps most often while diving. Look under ledges and in reef holes for their long, white antennae, often sticking out far enough to give away the shrimps' hiding place. Barber pole shrimps sometimes clean fish. This one probes a yellowmargin moray eel's face.

Conger Eel *(puhi ūhā), Conger* sp. Because conger eels are nocturnal, hiding in the reef during the day and then venturing out at night in search of food, these creatures are a rare find. Conger eels have strong teeth that they use to eat fish and sometimes shrimps, but these teeth aren't the long, sharp needles of many moray eels. Conger eels, which grow to about 4 feet long, are brown and gray.

Bluestripe Snapper *(ta'ape), Lutjanus kasmira.* You'll often see these yellow fish with blue stripes hanging together in large groups just above the reef. Bluestripe snappers, which commonly grow to 10 inches long, are not native to Hawai'i. The state introduced them from Tahiti in 1956 as a gamefish and they thrived. Snappers eat crabs, shrimps, and small fish.

Argus Grouper *(roi)*, *Cephalopholis argus.* This popular gamefish is not native to Hawai'i but was introduced here from French Polynesia in the 1950s. It has flourished. Argus groupers, which average about 5 pounds, are dark brown with light blue spots. Like other groupers, Argus groupers eat crabs, shrimps, and other fish.

Smalltail Wrasse, *Pseudojuloides cerasinus.* Male. These wrasses are usually found close to rocky bottoms at depths of 60 feet or more. Smalltail wrasses grow to about 5 inches long. Males are blue-green with a yellow stripe on each side.

Smalltail Wrasse, *Pseudojuloides cerasinus.* Female. Like most members of the wrasse family, male and female smalltail wrasses are different in both color and patterns. Some female wrasses, which are reddish in color, turn into males when males are in short supply.

Shortnose Wrasse,
Macropharyngodon geoffroy.
Compared with other members of the
family, these wrasses have blunt snouts,
hence their common name. Large teeth
in the throats of these wrasses help the
fish grind up snails, their most common
meal. Shortnose wrasses are yellowish
orange and covered with light blue
spots. These fish grow to about 6 inches long and live only in Hawaiian
waters.

Pavo Razorfish, *Xyrichtys pavo.*
Juvenile. This member of the wrasse
family lives in open sandy areas, diving
into the sand when threatened. Like
several other kinds of wrasses, juvenile
pavos look extremely different from
their adult forms. This young brown
fish holds its dangling, overhead fin forward, then drifts, looking like a dead
leaf. Adults lose this coloration and
unusual fin, eventually turning gray to
yellowish white.

Ringtail Wrasse *(po'ou), Cheilinus
unifasciatus.* To identify this wrasse,
look for a pink fish with a white ring
around its tail. These wrasses, which
grow to about 18 inches long, are fairly
common in water from 30 feet to
below scuba limits. Their most common food is fish, but they also eat crabs
and brittle stars.

Scrawled Filefish, *Aluterus scriptus.* If
you see this filefish, you probably won't
forget the sight. Scrawled filefish can
grow to 2 feet long and have long,
brushlike tails. Their common name
describes the blue, wavy marks on the
body. These fish are not abundant in
Hawai'i but are found in the deeper
waters of Hanauma Bay. Scrawled filefish
eat bottom-dwelling plants and animals.

Potter's Angelfish, *Centropyge potteri.* This fish is found only in Hawai'i and is the only common angelfish on Hawaiian reefs. Potter's angelfish, which grows to about 5 inches long, has a small home range and stays close to it, usually hiding in the coral at the approach of a diver. This blue, orange, and black-patterned fish eats mainly seaweed. It was named for the first director of the Waikiki Aquarium, Frederick Potter.

Redbar Hawkfish *(piliko'a), Cirrhitops fasciatus.* Like other hawkfish, the 5-inch-long redbar sits motionless on rock and coral waiting to ambush fish, crabs, and shrimps. Look for a red-and-black fish with white bars on the body and a black spot on each gill cover.

Blackside Hawkfish *(hilu piliko'a), Paracirrhites forsteri.* The common name of this lovely hawkfish doesn't do it justice, because only the rear, upper part of its body is black. Red spots on the head dissolve into orange, red, and white stripes on the body. Given their gaudy colors and patterns, blackside hawkfish are surprisingly successful at ambushing small passing fish during the day. These hawkfish grow to nearly 9 inches long.

Manta Ray *(hāhālua), Manta* sp.
The manta rays most often seen near
Hawai'i's shorelines measure up to 12
feet across. These unusual-shaped fish
may look frightening but are harmless
to people. Mantas swim forward, direct-
ing small drifting animals into their
mouths with a special lobe on each side
of the head. Look up to see the white
undersides of manta rays swimming at
the surface near the outside edge of the
reef. The backsides of these fish are
dark gray. Unlike Hawai'i's other rays,
manta rays have no tail barbs.

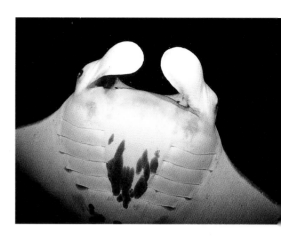

Manta Ray Mouth *(hāhālua), Manta*
sp. Strainers in the gill chambers of
mantas separate food from water.
The food goes to the stomach and the
water flows out the gills. As the water
exits, the gills swap carbon dioxide for
oxygen.

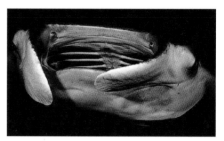

Whitetip Reef Shark, *Triaenodon obesus.* Unlike some other sharks, whitetips
can pump water over their gills, allowing the fish to breathe without swimming forward. Because of this characteristic, divers usually see whitetip reef sharks resting
under ledges and in caves. These sharks aren't aggressive toward people; however,
heed their family's reputation.It's a bad idea to disturb or chase any shark. Whitetips
usually eat octopuses, fish, and shellfish. These gray sharks with white fin markings
grow to about 6 feet long.

Speckled Scorpionfish, *Sebastapistes coniorta.* Scorpionfish get their common name from venomous spines that cause painful wounds if grabbed or stepped on. Speckled scorpionfish, which grow to only about 4 inches long, have small dark spots over their mottled brown bodies. Look for these fish hiding in antler coral from shallow water to about 80 feet deep. Speckled scorpionfish eat mostly crabs and shrimps.

Leaf Scorpionfish, *Taenianotus triacanthus.* The little leaf scorpionfish earned its name from its habit of rocking back and forth like a leaf in surging water. This fish has the unusual habit of occasionally shedding its outer layer of skin. Look carefully for these 4-inch-long fish camouflaged in combinations of red, yellow, brown, and black, from 2 feet to beyond scuba limits. Leaf scorpionfish eat crabs, shrimps, and fish.

Titan Scorpionfish *(nohu),* *Scorpaenopsis cacopsis.* These fish are found only in Hawaiian waters, where they grow to about 20 inches long. Titan scorpionfish can be hard to find even though these fish are large and fairly common at the outer edges of reefs, from 15 feet to beyond scuba limits. Look for a reddish orange fish holding perfectly still among matching coral rocks. Titan scorpionfish eat small fish.

Hawaiian Lionfish,
Dendrochirus barberi.
These 6-inch long fish are found only in Hawaiian waters, from about 3 feet deep to beyond scuba limits. Look for these scorpionfish family members drifting motionless in reef recesses, waiting for unsuspecting prey like fish, crabs, or shrimps to pass by. Hawaiian lionfish are reddish brown with dark spots on body and fins. Their venomous spines can cause painful wounds.

'Ewa Blenny, *Plagiotremus ewaensis.* This 4-inch-long fish hovers just above the reef waiting for larger fish to pass by. When this happens, the blenny strikes at speed, nipping off scales and mucus. 'Ewa (pronounced EH-va) blennies will also strike divers, but contact is usually just a light touch rather than a real bite. Look for these little orange fish with blue stripes at 30 feet and deeper. 'Ewa blennies are found only in Hawaiian waters.

Spotted Coral Blenny *(pāoʻo kauila), Exallias brevis.* Tentacles above the eyes of these 6-inch-long beauties give them a wide-eyed, innocent expression, while spots cover their white-and-red bodies. Female shortbodied blennies lay sticky, yellow eggs on rock surfaces; males then guard the eggs until they hatch. Here a male protects eggs, the yellow masses on the rocks just behind him. Spotted coral blennies eat the soft bodies of stony corals.

Commerson's Frogfish, *Antennarius commersonii.* Even the most alert divers don't often see frogfish because they look so much like rocks. Commerson's frogfish grows to about 12 inches long and can be yellow, red, brown, black, or all combinations of these colors. Each frogfish carries a built-in fishing pole on its snout with a lure hanging near the mouth. When a small fish comes close to inspect the lure, the frogfish snaps up an easy meal. Nine kinds of frogfish live in Hawai'i but none are common. Frogfish do not have venomous spines and are not dangerous to people.

Hawaiian Turkeyfish, *Pterois sphex.* Another Hawai'i-only native, this 9-inch-long red, white, and brown fish also hides under ledges and in reef holes. Turkeyfish hunt for shrimps, crabs, and other invertebrates at night. Like lionfish, turkeyfish also have venomous spines that can cause painful wounds.

Hawaiian Squirrelfish (*'ala'ihi*), *Sargocentron xantherythrum.* Look for these big-eyed, nocturnal fish in caves and other dark places in about 60 feet of water. It's common to see the red Hawaiian squirrelfish staring back at you from these hiding places. These 6- to 7-inch-long fish are found only in Hawaiian waters.

Squirrelfish (*'ala'ihi*), *Neoniphon aurolineatus.* Most members of the squirrelfish family hide in caves and under ledges during the day then venture out at night hunting crabs, shrimps, and other invertebrates. Unlike other family members, this 10-inch-long squirrelfish has yellow stripes on its body with red spots between them. Usually these fish are found at depths of 100 feet and deeper, but in Hanauma Bay you can find them in shallower water.

Spinner Dolphin, *Stenella longirostris.* Spinner dolphins, which grow to about 6 feet long, are rare inside the bay but sometimes swim near the outer edges. Spinners get their common name from their habit of leaping out of the water and spinning one or more times on their tails. Most spinner dolphins live far offshore, but Hawai'i has its own resident population. If you're lucky enough you see dolphins, stay as still as you can and let them come to you if they're so inclined. It's illegal to pursue these protected mammals, which eat fish and squid.

INDEX

ABOUT THE AUTHOR

Susan Scott earned a bachelor's degree in biology from the University of Hawaii in 1985 and is a graduate of the university's Marine Option Program. Scott writes a weekly column for the *Honolulu Star-Bulletin* called "Oceanwatch" and is a contributing editor for *Hawaii Magazine*. Her previous books are *Oceanwatcher: An Above-Water Guide to Hawaii's Marine Animals* and *Plants and Animals of Hawaii*. (Photograph by Patrick Ching)

ABOUT THE PHOTOGRAPHER

David R. Schrichte is a musician in the Navy and an award-winning marine photographer. His photographs have appeared in both local and national publications. Since 1985, Schrichte has worked tirelessly to protect marine animals, volunteering his time, energy, and photographs to promote this cause. He took all the photographs in this book with either his Canon F1 or Nikonis 3. (Photograph by Theresa Schrichte)